# WHY CLOUDS HAVE SHAPES TODAY

ISBN 0-89868-446-3–Library Bound
ISBN 0-89868-447-1–Soft Bound
ISBN 0-89868-448-X-Trade

A PREDICTABLE WORD BOOK

# WHY CLOUDS HAVE SHAPES TODAY

Story by Janie Spaht Gill, Ph.D.
Illustrations by Lori Wing

ARO PUBLISHING

4

Once there were only flat clouds
that stretched from east to west.

Then, out of them was born a cloud,
much different than the rest.

Little Cloud was smaller
and sat lower in the sky.

Each day his shape would change,
as through him planes would fly.

7

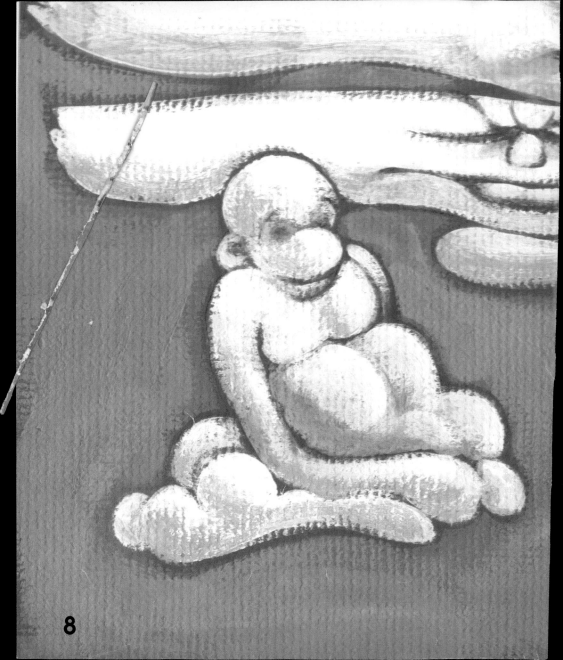

8

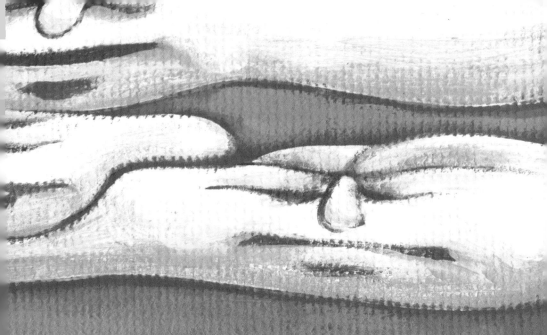

The white, flat clouds above him,
would laugh at shapes he'd make.

"What a silly cloud you are!
Why can't you stay one shape?"

9

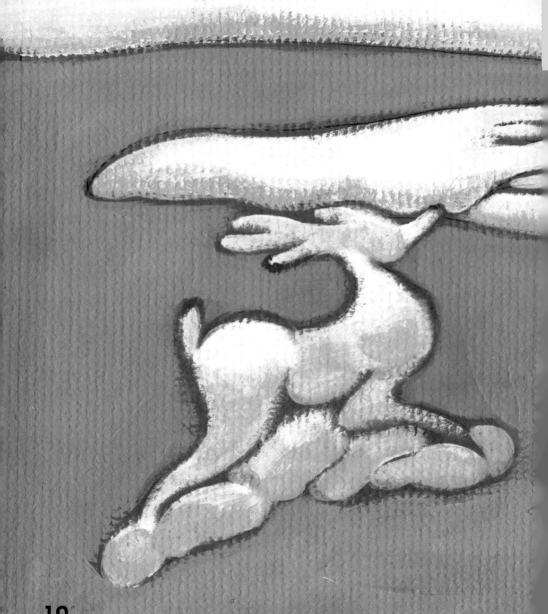

10

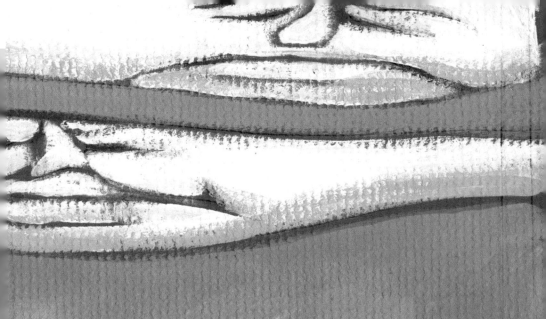

First he became a reindeer,

12

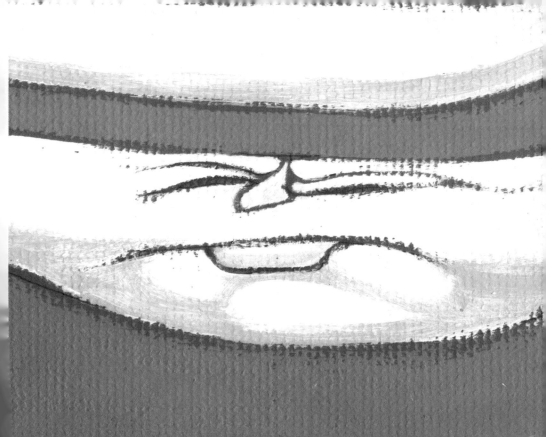

then he became a duck.

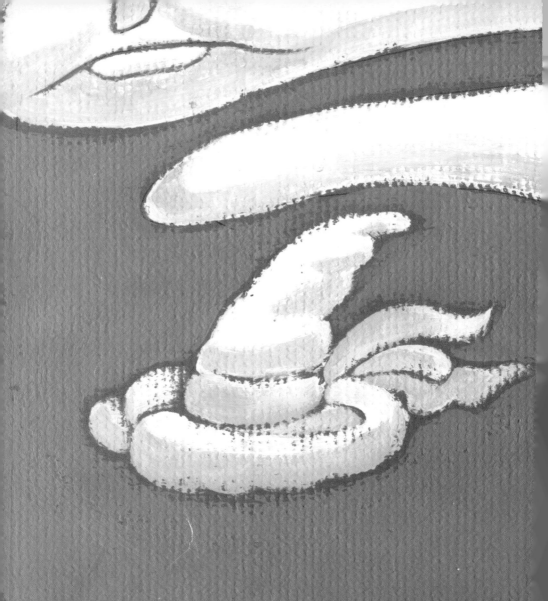

14

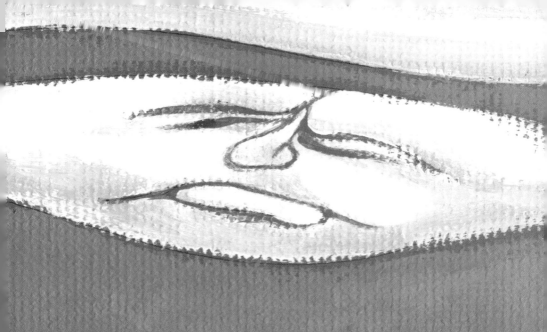

Then he'd change into a witch's hat,

16

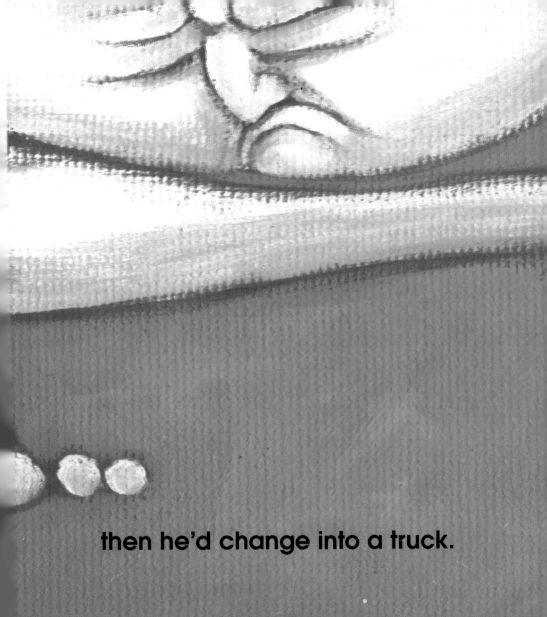

then he'd change into a truck.

18

The people on the ground
would point to him and say,

"Oh Little Cloud, it's so much fun
to guess your shape each day!"

The white, flat clouds above him,
were jealous at the fuss.

They thought, "We're never noticed,
so change our shape we must."

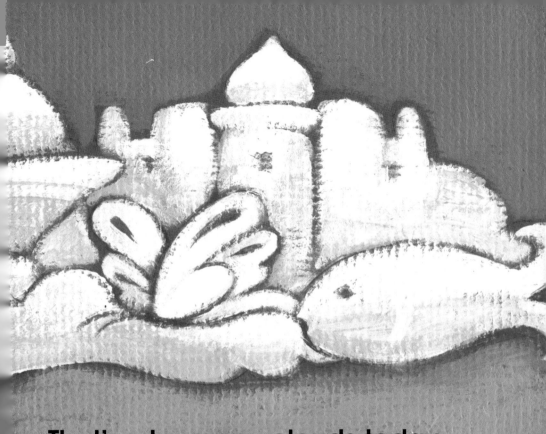

That's why many clouds today,
form shapes up in the sky.

They want you each to guess their
shape, as they go floating by.

23

| DATE DUE | | | |
|---|---|---|---|
| | | | |
| | | | |
| | | | |
| | | | |
| | | | |
| | | | |
| | | | |

HPARX    +
*Friends of the*    E
*Houston Public Library* GILL

GILL, JANIE SPAHT
WHY CLOUDS HAVE
SHAPES TODAY